# When a Tree Falls

For my tree child, Max, who carries
the forest in his heart.
For all the people working to protect
our precious old-growth trees.
—K. P.

To my parents and Paul. Thank you for your
unconditional love and support.
—E. B.



Library of Congress Cataloging-in-Publication Data available.

ISBN 978-1-7972-1867-0

Manufactured in China.

Design by Ryan Hayes.
Typeset in Morion.
The illustrations in this book were rendered digitally.

10 9 8 7 6 5 4 3 2

Chronicle Books LLC
680 Second Street
San Francisco, California 94107

Chronicle Books—we see things differently.
Become part of our community at www.chroniclekids.com.

# When a Tree Falls

## Nurse Logs and Their Incredible Forest Power

By
Kirsten Pendreigh

Illustrated by
Elke Boschinger

CHRONICLE BOOKS
SAN FRANCISCO

A tall tree suns and sways.

She is a place to grow,

to rest,

to shelter.

A lookout,

an aerie,

a roost,

a roof.

Years pass.

Sapsuckers drill,
hidey-holes fill,
bear cubs rub
and claw.

Woodpeckers peck.

The tree grows old.

Sap slows.

Roots let go.

Green goes brown.

The tree falls

down,

down—

*creeeeeeak!*

*Crassshhhh!*

*Thud!*

Creatures skitter,

scatter.

Crouch.

Watch.

And wait.

The tree is still.

No more sunning

or swaying.

Her new life begins

as Nurse Log.

Creatures return.

Coyote sniffs,

chipmunks *chit-chitter*,

towhees tiptoe,

bobcat scrapes.

Days pass.

Slugs slither.

Moss creeps.

Conks clasp.

Lichen tickles.

Jelly fungus sticks.

Nurse Log welcomes them all.

She is a place to grow,

to rest,

to shelter.

On cold days

and rainy nights,

Nurse Log warms

her friends.

Big.

Small.

Smaller.

And smallest.

Nurse Log holds the rain
in her heart
for the hot summer,
when the forest dries.

Then she shares

her precious damp

with the ferns,

with the mosses,

with a seed set down by the wind.

Nurse Log feeds that seed

to seedling,

to sapling.

All the while,

bees hive,

spiders spin,

mushrooms mushroom.

Squirrels stash

and scratch,

weave nests.

A small human sits and rests.

Nurse Log mothers them all.

She is a lap,

a bed,

a cradle.

She lifts to light.

Her tree child stretches up,

branches out,

roots down,

hugs his nurse mother.

Years pass.

Seasons come

and go.

Creatures grow.

Nurse Log dwindles,

crumbles.

She returns to earth.

Now her tall child

suns and sways.

He is a place to grow,

to rest,

to shelter.

Just like his mother.

## What are nurse logs?

Nurse logs are fallen trees that provide shelter, food, and water for all kinds of animals, from small, crawly centipedes to big, furry bears. They also nourish plant life, from tiny, fuzzy fungi to green tree seedlings that grow hundreds of feet tall. Nurse logs improve seedlings' chances of survival by giving them better access to light, water, and nourishment. *To nurse* means to nurture and care for something. That's why these logs are called *nurse logs*.

## How do nurse logs nurture new life?

After trees fall, microbes (tiny living organisms that include bacteria and fungi) get to work. Just as you chew on a hard piece of toast to soften it, these little microbes chew on the hard wood. Then insects like beetles and termites arrive and begin munching and burrowing. The holes they make allow water to penetrate the bark and soften the wood even more. The decay begins! And guess who loves eating insects? Birds! Especially woodpeckers, who start poking even more holes in the wood. Now it's easier for little plant seedlings to push their roots inside. And, lucky for them, the microbes already living in the nurse logs release nutrients that help the seedlings grow.

Squirrels, chipmunks, and other small animals make nests in the hollows that the woodpeckers leave behind. Snakes and frogs shelter under nurse logs. And if the root hollows are big enough, bears make their dens in nurse logs! Between 20 and 40 percent of all known animal and plant species depend on deadwood for survival.

## How do nurse logs help new trees grow?

Imagine that you are a tiny tree seed trying to grow in the dense forest. What does a tree need to grow? Just like you, it needs food, water, a safe home, and sunshine!

When a seed lands on a nurse log, it has a much better chance of survival.

- **It has found a water source.** To germinate, the most important thing a seed needs is moisture. Once the seed sprouts and becomes a seedling, it still needs water, or its tiny roots will die. In summer, the forest floor can dry out, but the dense, rotting wood of a nurse log holds lots of moisture, like a sponge, even on the hottest days. A nurse log can have up to 20 times the moisture content of nearby soil.
- **It has found a way to get more light.** Once the little seed becomes a seedling, it needs sunlight! Up on a nurse log platform, it doesn't have to compete as fiercely with the ferns, shrubs, and other tree seedlings already densely growing on the shady forest floor.
- **It has found protection from deer.** In many forests, seedlings are eaten by hungry deer. If a seed lands on a big, wide nurse log, like the ones found in the Pacific Northwest, it is harder for deer to reach. Deer also prefer browsing in open areas—they don't like trying to move around big logs and fallen branches. The more nurse logs, the bumpier the ground becomes, and the less deer want to visit!

- **It has found some protection from disease.** Some forest soils have pathogens that can infect tender seedlings. These pathogens can't live on deadwood like a nurse log.
- **It has found a source of food.** The decaying wood of a nurse log has some nutrients and natural fertilizers, like phosphorus and nitrogen, that feed young seedlings. Leaves and animal droppings that accumulate on top of a nurse log also become fertilizer. But as a seedling gets bigger, it needs more food, so it reaches its roots down to the forest soil, where more minerals and nutrients are available. Often, the roots wrap around the nurse log like a hug.

## How do nurse logs keep the whole forest healthy?

- **Nurse logs help the forest floor**. They add organic matter and moisture to the soil. In forests where dead trees are left to rot, the soil is loamier and full of healthy microbes. Nurse logs also help prevent soil compaction and soil erosion on slopes.
- **Nurse logs help other plants**. By discouraging deer, they give plants growing nearby a better chance to survive.
- **Nurse logs help rivers and fish**. Trees that fall into forest streams create natural dams and calm pools. These prevent the riverbanks from eroding and provide a place for salmon to rest and hide their eggs. Then those salmon feed other animals in the forest!

Forest managers used to remove dead trees for their timber. They believed it was better to keep forests "tidy." Now, to have a healthy forest ecosystem that benefits all creatures, forest ecologists try to leave dead trees where they fall. Unlike a messy bedroom, a messy forest is a wonderful thing!

## Are all nurse logs fallen trees?

While the nurse log in this story came from a large tree that toppled over, trees can also snap higher up during storms, leaving behind jagged trunks (known as snags) that nurture life, as well. Single branches might fall off, becoming smaller nurse logs on the ground. Nurse logs are also made when humans cut down trees, leaving behind flat stumps. Take a closer look the next time you are in the woods, in a park, or even in a backyard—nurse logs are everywhere!

## The trees in this story

The first tree in this story is a big, old-growth Douglas fir in the rainforest of the Pacific Northwest. Douglas firs can live for more than 1,000 years and can grow more than 300 feet (91.4 metres) tall and 10 feet (3 metres) wide! But when they are stressed by environmental factors like drought or by disease, they die younger. Woodpeckers and other animals can damage bark and introduce disease. Sometimes fungal infections cause tree roots to rot and decay. Without strong roots holding them up, trees easily topple over, often in windstorms or after heavy rains wash away the soil beneath them. Sometimes trees die when they are struck by lightning or burned in fires. When they die, Douglas firs give back to the forest by becoming nurse logs.

The second tree, the little seed that grows in the nurse log, is a western hemlock. Hemlocks can grow up to 150 feet (45.7 metres) tall and 9 feet (2.7 metres) wide, and most sprout in deadwood! Young hemlocks fairly quickly need more food than their nurse logs can provide, so they reach their roots down to the forest floor. As they get taller, their nurse logs get smaller and finally decompose into soft, loamy soil. Hemlocks often have hollows under their roots where animals can shelter. That's the "memory" of their nurse logs!

The way Douglas firs nurse other species like western hemlocks shows how connected all species are and how we all need to help one another if we want healthy forests and a healthy planet.